discovermore

Exploring Physical Science

Magnetism

Marie Harts

Britannica® Educational Publishing
IN ASSOCIATION WITH
THE ROSEN PUBLISHING GROUP

Published in 2024 by The Rosen Publishing Group, Inc.
2544 Clinton Street, Buffalo, NY 14224

Editor: Shannon Harts
Book Design: Rachel Rising

Photo Credits: Cover 279photo Studio/Shutterstock.com; (series background) Dai Yim/Shutterstock.com; p. 4 Rich Carey/Shutterstock.com; p. 5 Andrey_Popov/Shutterstock.com; p. 6 Vasin Lee/Shutterstock.com; p. 7 guruXOX/Shutterstock.com; p. 9 Slatan/Shutterstock.com; p. 9 alphaspirit.it/Shutterstock.com; p. 10 Maximillian cabinet/Shutterstock.com; p.11 https://commons.wikimedia.org/wiki/File:Amundsen-scott-south_pole_station_2007.jpg; p. 12 Elena11/Shutterstock.com; p. 13 Artsiom P/Shutterstock.com; p. 14 Canan Asik/Shutterstock.com; p. 15 Pressmaster/Shutterstock.com; p.16. https://commons.wikimedia.org/wiki/File:De_magnete_magneticisqve_corporibvs_et_de_magno_magnete_tellure_physiologia_1600_(46162682).jpg; p.17 Natalia Varlei/Shutterstock.com; p. 18 Dragana Gordic/Shutterstock.com; p. 19 PixieMe/Shutterstock.com; p. 20 vvoe/Shutterstock.com; p. 21 Pixel-Shot/Shutterstock.com; p. 23 buffaloboy/Shutterstock.com; p. 23 nadia_if/Shutterstock.com; p. 24 Surasak_Photo/Shutterstock.com; p. 25 Funtay/Shutterstock.com; p. 26 zlikovec/Shutterstock.com; p. 27 YAKOBCHUK VIACHESLAV/Shutterstock.com; p. 28 Den Rozhnovsky/Shutterstock.com; p. 29 tetiana_u/Shutterstock.com.

Cataloging-in-Publication Data

Names: Harts, Marie.
Title: Magnetism / Marie Harts.
Description: New York : Britannica Educational Publishing, in Association with Rosen Educational Services. 2024. | Series: Discover more: exploring physical science | Includes glossary and index.
Identifiers: ISBN 9781641900546 (library bound) | ISBN 9781641900539 (pbk) | ISBN 9781641900553 (ebook)
Subjects: LCSH: Magnetism--Juvenile literature.
Classification: LCC QC753.7 H37 2024 | DDC 538--dc23

Manufactured in the United States of America

Some of the images in this book illustrate individuals who are models. The depictions do not imply actual situations or events.

CPSIA Compliance Information: Batch #CWBRIT24. For further information contact Rosen Publishing at 1-800-237-9932.

Contents

What Is Magnetism?

What force allows trains to speed down tracks at 375 miles (603.5 km) per hour, people to find their way in the woods, and many everyday machines to work? The answer is magnetism.

Magnetism is a natural force that can either bring certain metal objects together, like a magnet and a refrigerator door, or keep them apart. Some metals have this power naturally.

Studies have shown that many animals, including sea turtles, can sense Earth's magnetic field and use it to migrate.

Man-made magnets are created from a group of metals called ferromagnetic metals. Iron, nickel, and cobalt are examples. These metals can be made magnetic by being exposed to a magnetic field.

Magnetism is an invisible force—it can be felt, but not seen. Naturally magnetic rocks are found on Earth, and Earth itself is one giant magnet! Scientists are working to learn more about Earth's magnetism, such as how it can help animals **navigate**.

WORD WISE

Navigate means finding the right direction for travel.

Making a Magnet

An atom is a tiny particle of matter. Atoms make up everything on Earth. Inside atoms are protons, neutrons, and **electrons**. When electrons spin around the center of the atom, tiny magnetic forces are created. If most or all of the electrons spin in the same direction, the magnetic forces join together to make a magnet.

Lightning can turn some rocks into magnets because it is a fast-moving electric current with a magnetic field.

A refrigerator magnet is made by heating and mixing iron rust and plastic or rubber. This mixture is then poured into a mold and exposed to a strong magnetic field.

Magnets can be made by rubbing a piece of ferromagnetic metal in one direction against an existing magnet many times. A magnetic force is created when the metal's spinning electrons are all pulled in one direction by the existing magnet.

Consider This

Objects can lose their magnetic abilities by heating them above their Curie temperature, which is the heat at which an object loses its ability to be magnetized. Why do you think high temperature causes a metal to lose its magnetic power?

WORD WISE

An electron is one of the three particles that make up atoms. Electrons circle an atom's nucleus, or core, which includes positively charged protons and neutrons with a neutral charge.

Temporary and Permanent Magnets

Magnetic fields can be found everywhere on Earth, and throughout the universe! Even objects such as wood or plastic have some magnetism, but it's very weak.

Some magnets have electrons that move in one direction no matter what's happening around them. These are called hard, or permanent, magnets. It's difficult to change the magnetism of hard magnets.

Other materials, such as iron, are called soft, or temporary, magnets. They are attracted to hard magnets, but are not themselves permanent magnets. Temporary magnets must be in the presence of permanent magnets to act like magnets.

Many magnets are shaped like horseshoes because having two poles, or ends, of the magnet next to each other creates more pull force.

compare and contrast

Iron nails are temporary magnets. They may seem to be magnetic in the presence of a horseshoe magnet, but what happens when you try to use the iron nail as a magnet without the horseshoe magnet?

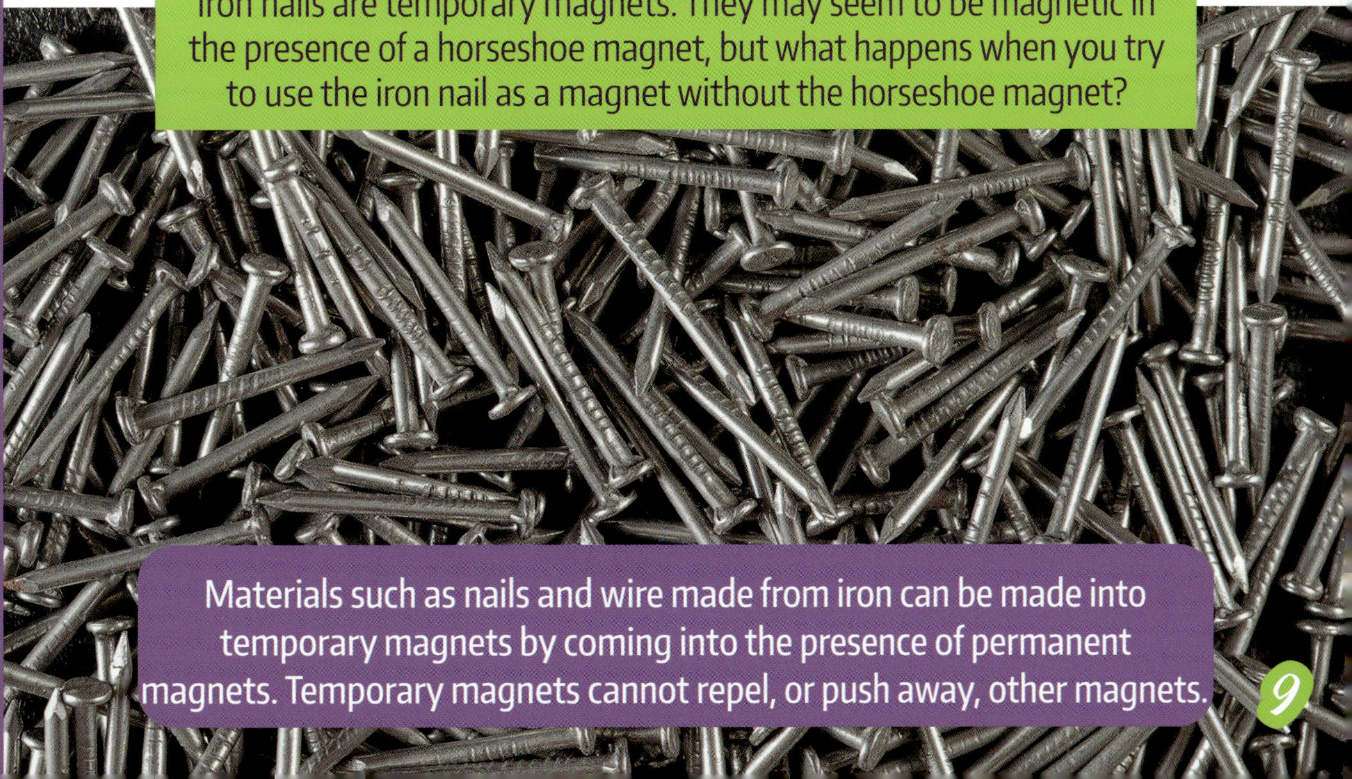

Materials such as nails and wire made from iron can be made into temporary magnets by coming into the presence of permanent magnets. Temporary magnets cannot repel, or push away, other magnets.

Two Poles

Any magnet you can find has two ends called poles. One end is called the north pole and the opposite end is called the south pole. If opposite poles are held near each other, they will come together. However, holding like poles near each other makes them repel one another.

Earth's geographic North Pole (shown above) is in the middle of the Arctic Ocean. Earth's magnetic North Pole is found, for now, on Ellesmere Island in Northern Canada. However, Earth's magnetic North Pole is shifting.

Earth's geographic South Pole (shown here) is in Antarctica, about 300 miles (480 km) south of the Ross Ice Shelf. Its magnetic South Pole is shifting, but found for now on the Adélie Coast in Eastern Antarctica.

Earth has two poles as well since it's a large but weak magnet. Earth's magnetic North and South Poles are where its magnetism is strongest. Earth's magnetic field is weakest at the equator, which is an imaginary line around Earth that is the same distance from the North and South Poles.

Consider This

Earth has two sets of poles. Earth's geographic North and South Poles are where its lines of longitude (meridians) meet. Earth's magnetic North and South Poles are positions where its magnetic field is vertical (up and down). Do you think a compass points to the geographic or magnetic North Pole?

The Magic of Magnetic Fields

Can you move something without touching it? A magnetic force can! Magnetic forces work through air, water, and even nonmagnetic objects. For example, if you put a piece of paper between two magnets, they will still act as if nothing separated them. The area around a magnet that has the magnetic force is called a magnetic field.

Earth's magnetic field acts like a shield, keeping it safe from charged particles that stream from the sun.

Mars and Venus don't have a strong magnetic field like Earth and other planets do. Without its magnetic field, Earth would be more like Mars!

A compass uses Earth's magnetic field to show direction. One end of the magnet in a compass is attracted to the magnetic North Pole. The magnet is mounted so that it is free to turn. A compass can help you find your way because no matter what direction you are facing, one end of the compass will point towards the magnetic North Pole.

Consider This

Earth's molten outer core, where flowing iron creates electric currents, produces its magnetic field. Due to movement in the outer core, Earth's magnetic field is always shifting. Do you think this affects the direction in which compasses point?

Energy Creation Partners

Electricity and magnetism are both created by the movement of electrons. Electricity is simply the free movement of electrons, which creates an electric current. A current is a stream of charged particles. An electric current flowing through wires creates magnetic fields. An electric current can be used to create magnetism, and magnetism can be used to create an electric current. They are partners in the creation of energy.

An induction cooktop heats food using an electromagnetic field. Turning the cooktop on causes an electric current to travel through a coil of metal. This creates a magnetic field.

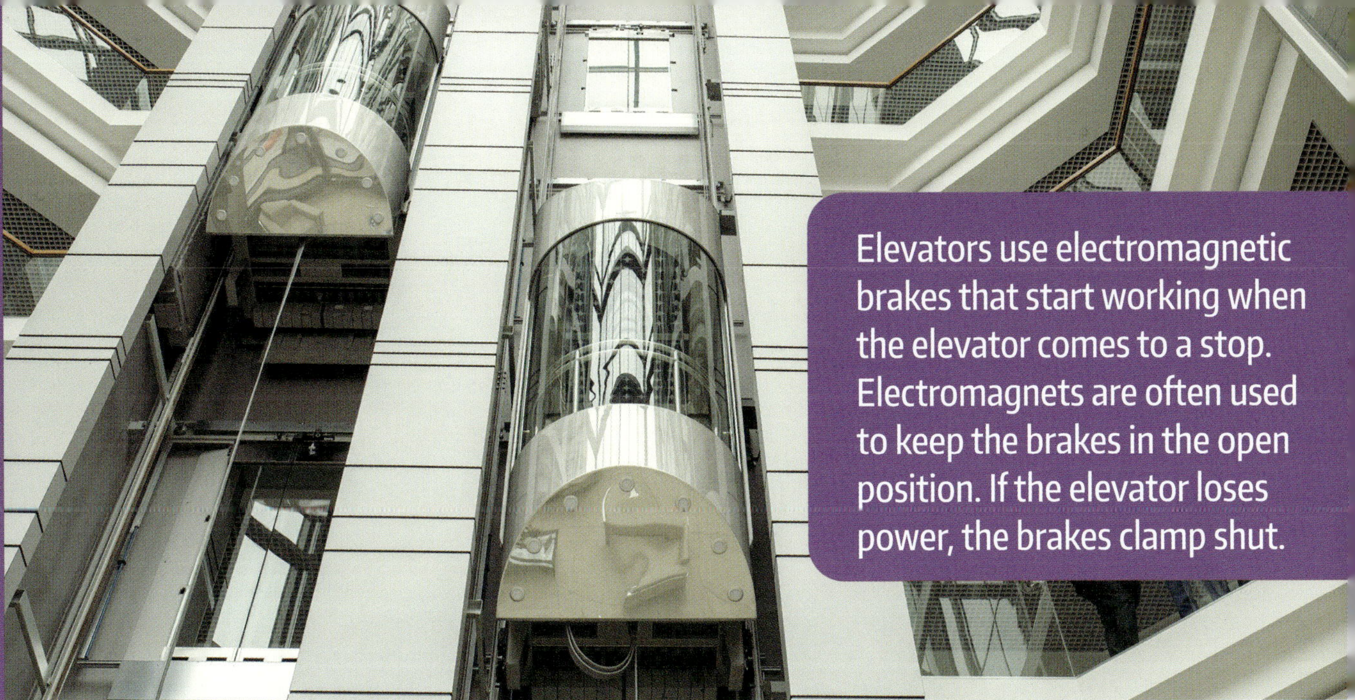

Elevators use electromagnetic brakes that start working when the elevator comes to a stop. Electromagnets are often used to keep the brakes in the open position. If the elevator loses power, the brakes clamp shut.

Applying electricity to a wire coiled around an iron rod creates an electromagnet. These are types of soft magnets that can be turned on and off by controlling the electric current. Electric engines, **generators**, and many consumer electronics are among everyday machines that use electromagnets.

WORD WISE

A generator is a machine that changes a certain form of energy, such as water, steam, or gasoline, into electricity. An electromagnetic generator has an electromagnetic shaft called a rotor surrounding coils of wires. When the rotor is turned, an electric current flows through the wire coils.

Discovering Magnets

Magnets are found in nature, meaning they weren't created by people. The ancient Greeks and Chinese discovered the first natural magnets, called lodestones. After rubbing metal on these stones, the metal would itself become magnetic. Lodestones were used for religious and spiritual ceremonies for thousands of years. The Chinese began using them as compasses for navigation about one thousand years ago.

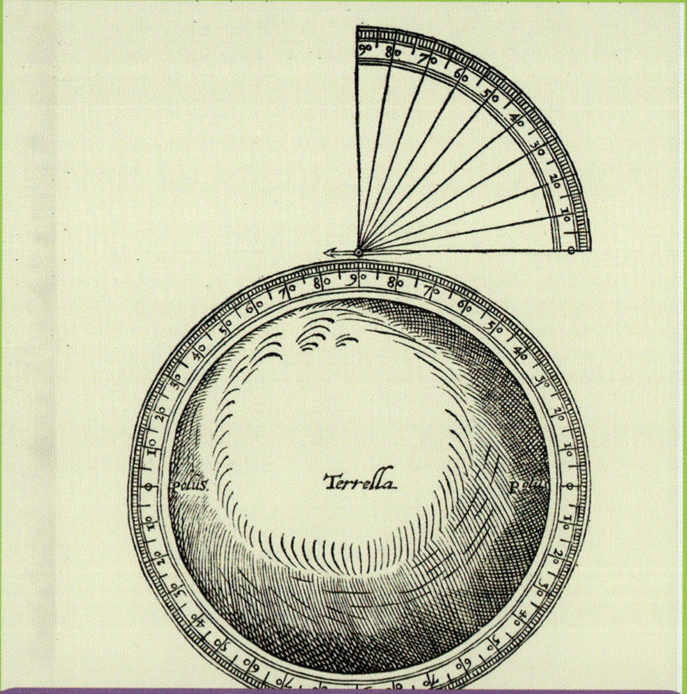

William Gilbert was able to show, using his versorium instrument similar to the instrument shown above, that Earth is a magnet like lodestone.

Danish **physicist** and chemist Hans Christian Ørsted was one of the first scientists to discover a connection between magnetism and electricity. He noticed a wire's electric current could deflect a magnetized compass needle.

Many centuries passed before people began to understand why a compass needle always pointed north. William Gilbert, a doctor from England, was the first to prove that Earth itself is a giant magnet. Around 1600, Gilbert made this discovery using a device called a versorium. This instrument includes a freely turning metal needle and a round lodestone called terrella.

Consider This

Before Gilbert discovered Earth's magnetic field, sailors used to think that compasses pointed north due to the North Star. In what other ways do you think Gilbert's discovery changed navigation and people's understanding of the world?

WORD WISE

A physicist is someone who works to understand the laws and properties that control energy, time, space, and matter.

Magnets Are Everywhere

In modern life, magnets are nearly impossible to avoid. Refrigerator doors have magnets in them to keep the doors closed and sealed. The headphones that you use to listen to music or to play video games also have magnets in them.

Magnets are a part of electric motors used in many household tools and appliances. The spinning parts of appliances such as blenders, mixers, washers, and dryers also often use magnets to get work done.

Magnets in speakers are used to turn electric audio signals into sound waves.

Hard drives are an important part of most computers and they include multiple magnets.

Consider This

Do you think the size of a magnet affects its strength? Why or why not? You can test this by comparing the strength of a large magnet to that of a small magnet.

WORD WISE

A hard drive is a device that stores a computer's operating system, applications (apps), and data files such as pictures and music. Disks called platters that are made of magnetic material and store information, or data, are an important part of hard drives.

Electronics and Magnets

You may not realize it, but nearly every time you use an electronic device, you're using at least one magnet. Electromagnets that turn an electrical current on and off are used in switches, such as those used for doorbells. They are also used in home security systems. When two magnets are separated by opening a window or door, the magnetic force is broken, causing the alarm to go off.

Microwave ovens use magnets both to make the spinning tray work and to create the microwaves that heat food.

Laptop computers can have magnets that sense when they are open or closed.

Smartphones could not vibrate without the use of an electromagnet. Computers also have magnets in fans that prevent them from overheating. The motors in these fans use magnetic fields to create the force needed to spin.

WORD WISE

Microwave ovens heat food using microwaves. Microwaves are a form of electromagnetic radiation similar to radio waves. Radiation is energy in the form of waves or rays that can't be seen.

Movement and Magnetism

If you're traveling somewhere, magnets will likely help you get there! In newer hybrid and electric cars, hard magnets are used to create needed power, while keeping the engine small and lightweight.

Modern roller coasters use magnets to create a thrilling ride. At the beginning of the ride, magnets repel each other for the initial, or first, push off. During the ride, magnets are used to slow down and speed up the cars. Even the brakes used to stop the cars at the end of the ride are magnetic.

Maglev, or magnetic levitation, trains are used for faster long-distance travel. Instead of riding on a track, these trains use magnetism to hover over the track. The fastest Maglev train as of 2021 is Japan's L0 Series, which reached a record speed of 375 miles (603.5 km) per hour.

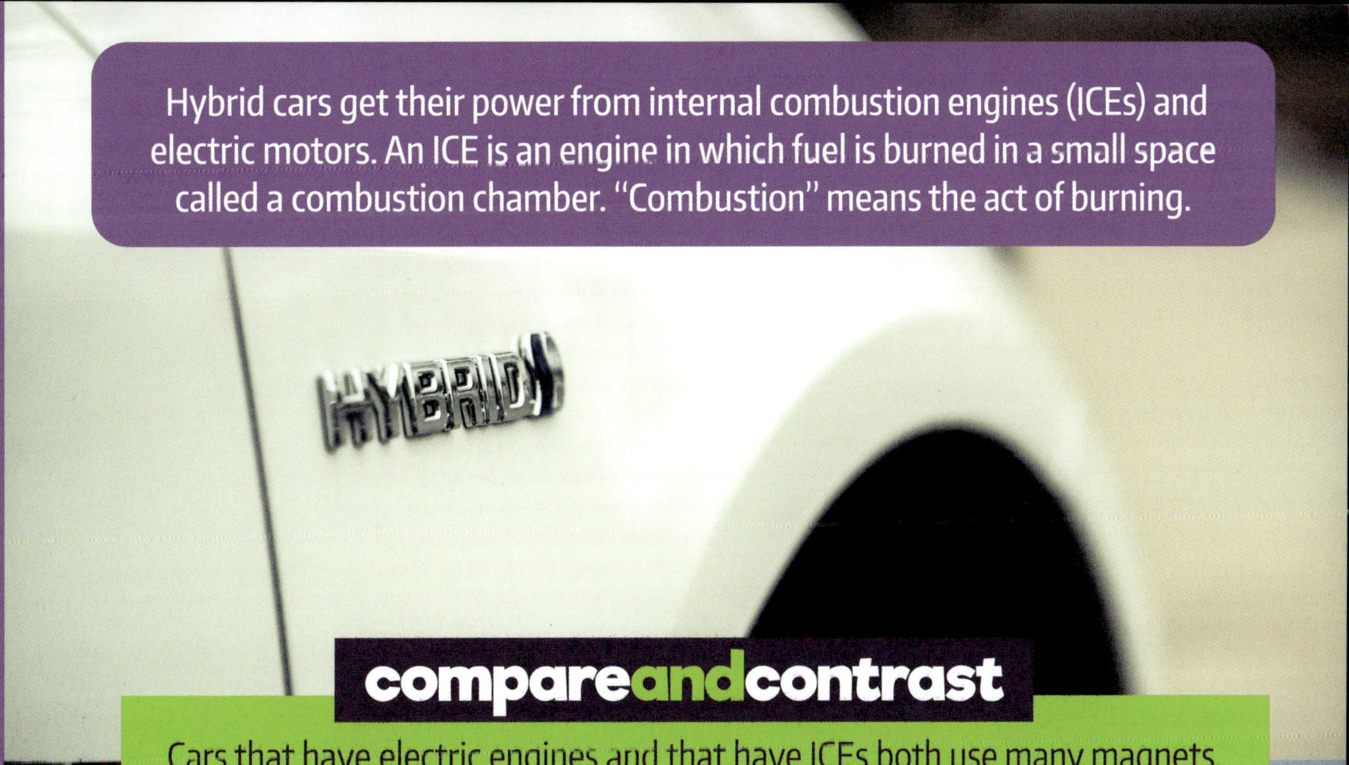

Hybrid cars get their power from internal combustion engines (ICEs) and electric motors. An ICE is an engine in which fuel is burned in a small space called a combustion chamber. "Combustion" means the act of burning.

compare and contrast

Cars that have electric engines and that have ICEs both use many magnets. Do you think magnets are more important for electric or ICE cars?

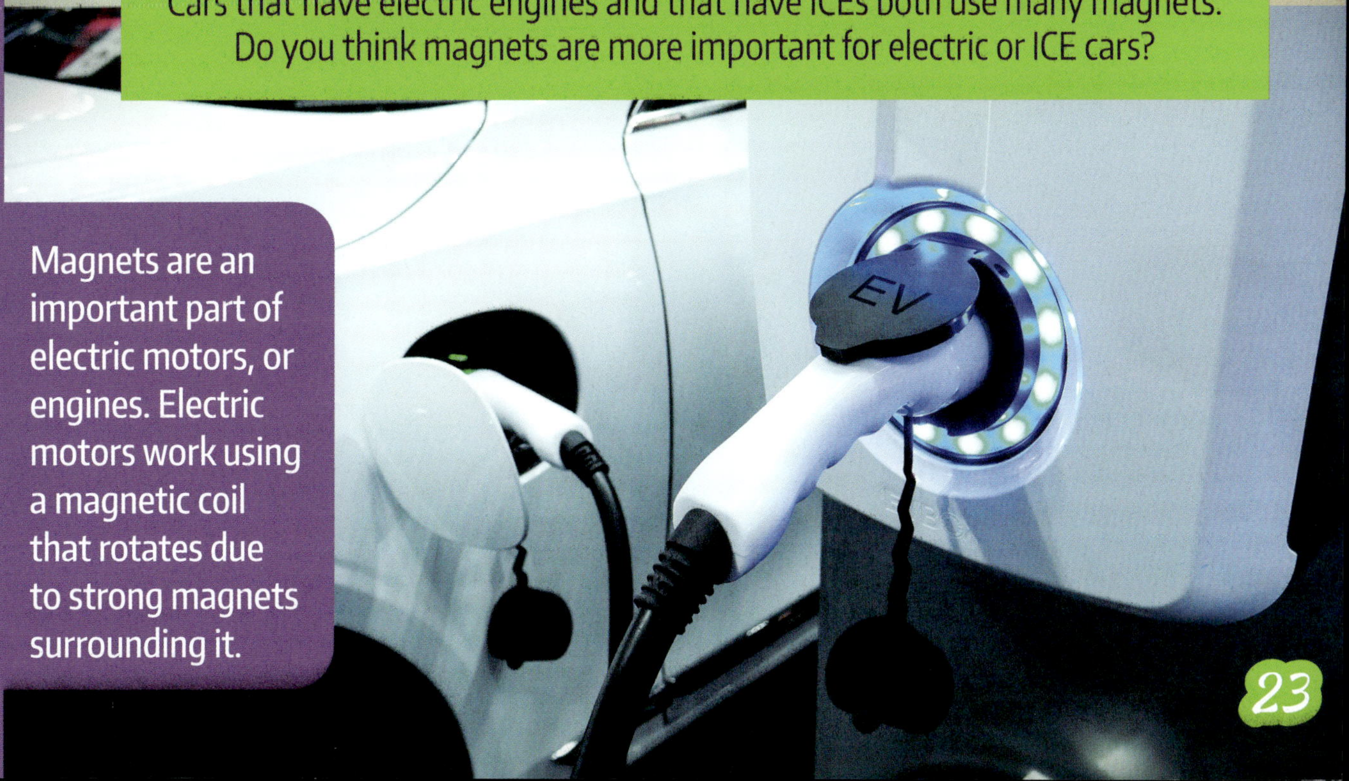

Magnets are an important part of electric motors, or engines. Electric motors work using a magnetic coil that rotates due to strong magnets surrounding it.

Magnets for Making Things

Machines that use magnets are commonly found on construction sites. Large electromagnets mounted on cranes attract heavy pieces of metal and lift them up. The crane then moves the metal to the place it needs to go. When the electromagnet is turned off, the metal piece is dropped in place.

Magnetic separators such as this are used in many industries, including the production of plastics, powders, and foods.

Magnetic drills such as this are often used in construction to drill holes in steel and other hard metals. They have electromagnets at their base that keep them steadily attached to steel.

Magnets can also be used to pick up small pieces of metal, like nails, from the ground to prevent workers from getting hurt and machines from being harmed. In a factory, magnets might be used to hold screws or nails in place to be tightened by machines. Magnets are also often used on assembly lines. For example, on an assembly line that makes cars, magnets are used to lift heavy parts such as the engine, outer shell, and doors.

Consider This

In food production, magnetic separators are often used to remove ferromagnetic materials. Why do you think this is important?

Medical Magnetism

Magnets have been used to relieve pain for hundreds of years. Although some people believe that magnets can cure painful illnesses, scientists have not been able to prove it. However, some studies have shown electromagnets can help with pain relief.

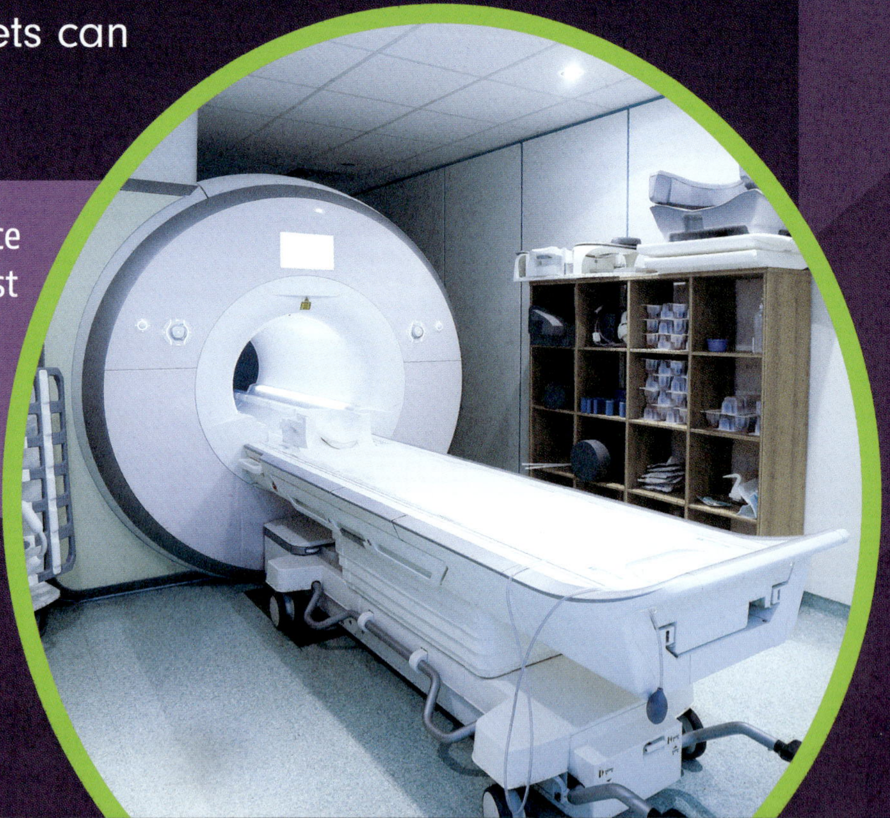

MRI machines can produce detailed pictures of almost every internal structure of the body using a strong magnet and radio waves.

Magnetic therapy is thought to work using the body's natural magnetic and electric fields, realigning the fields when they are out of balance. There's no current scientific proof magnetic therapy works, though.

Magnets do have a very important proven use in medicine, though. Magnetic resonance imaging, or MRI, machines create a magnetic field around a person. This magnetic field for a time aligns water **molecules** in a person's body. The machine then makes radio waves that cause aligned atoms to create faint signals. These signals are used to make pictures that allow doctors to see the inside of people's bodies.

WORD WISE

A molecule is two or more atoms grouped together into the smallest unit of a substance.

Magnets in Action!

You can experiment with magnets in your own home! Try these projects using magnets and household items to put your understanding of magnets to the test.

Turn a Needle into a Magnet

For this experiment, you will need a strong magnet and a sewing needle. Hold the needle by its eye. Take the magnet and brush it along the length of the needle in one direction about 30 times. Your needle is now magnetic! Test it by holding it near other metal objects, and see if you can feel the attraction.

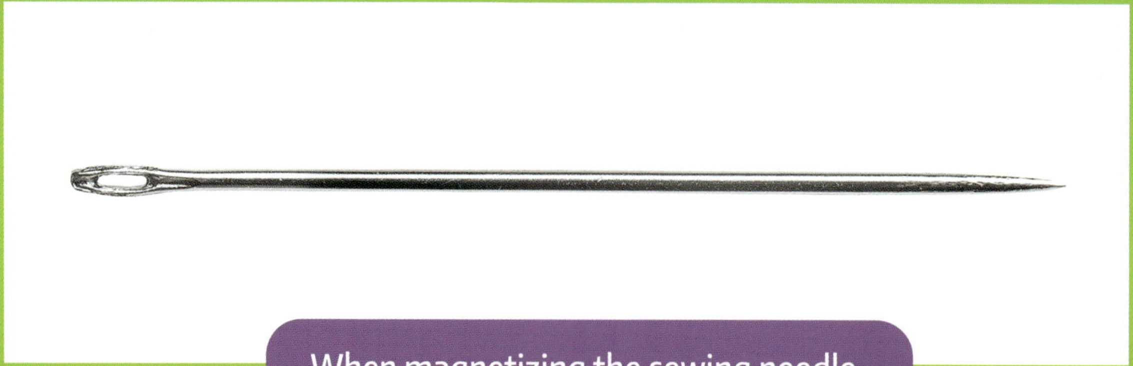

When magnetizing the sewing needle, be careful of its sharp tip!

There are multiple uses of a magnet's ability to work through glass—such as cleaning two sides of windows or fish tanks at once! Can you think of any other uses?

The Paper Clip Challenge

This experiment requires a magnet, a paper clip (or a nail), and a clear glass of water. Drop the paper clip into the glass of water. See if you can use the magnet on the outside of the glass to move the nail or paper clip. Does the magnet's magnetic force work through the glass? And if so, are you able to move the nail or paper clip out of the water without using your hands?

Consider This

Magnetic materials aren't just found inside your home. With an adult's okay, try going outside with a magnet and seeing what magnetic materials you can find.

Glossary

appliance A machine such as a stove, microwave, or dishwasher that does a certain job in people's homes and is powered by electricity.

assembly line A moving pathway in a factory where goods are put together piece by piece.

attract To pull toward; the opposite of "repel."

ceremony An event to honor or celebrate something.

chemist Someone who is trained in chemistry, which is a branch of science that involves the makeup, structure, and properties of substances, in addition to the changes they can undergo.

compass A device that uses a magnet to tell direction.

current The flow of electricity.

data Facts and figures.

experiment A test of a scientific idea. To experiment is to try out a new activity, idea, or way of doing something.

expose To cause something to be affected or influenced by something else.

geographic Relating to the physical location of a thing.

levitation The act of rising above the ground.

longitude The imaginary lines that run north and south to the left and right of the Prime Meridian.

migrate To move from one area to another for feeding or having babies.

molten Changed into a liquid form by heat.

operating system A computer's main program that controls its memory and processes, in addition to allowing a user to communicate with the computer.

religious Having to do with religion, which is a belief in and way of honoring a god or gods.

rotate To turn around a fixed point.

thrilling Causing excitement and pleasure.

turbine An engine whose central driving shaft includes a series of blades that are spun around by the force of water, steam, or air.

universe The whole world, space, and things people experience.

For More Information

Books

Dufresne, Emilie. *Magnets*. New York, NY: KidHaven Publishing, 2022.

Gaertner, Meg. *Magnets*. Minneapolis, MN: Cody Koala, 2020.

O'Daly, Anne. *Amazing Activities with Electricity and Magnetism.* New York, NY: Enslow Publishing, 2023.

Websites

40 Fun Magnet Experiments and Play Ideas
teachbesideme.com/magnet-teaching-play-ideas/
Check out more fun activities to learn about magnets.

How Are Magnets Used?
wonderopolis.org/wonder/how-are-magnets-used
Learn more about the many uses of magnets here!

Magnet and Magnetism
kids.britannica.com/kids/article/Magnet-and-Magnetism/353411
Find more information, including a short video, about magnets and magnetism.

Publisher's note to educators and parents: Our editors have carefully reviewed these websites to ensure that they are suitable for students. Many websites change frequently, however, and we cannot guarantee that a site's future contents will continue to meet our high standards of quality and educational value. Be advised that students should be closely supervised whenever they access the internet.

Index